Mathematical Olympia

for

Beginning Students 1

The Zeroth Book for Elementary Schoolers

(Workbook)

Educational Collection *Magna-Scientia*

The Zeroth Book for Elementary Schoolers

Mathematical Olympiads *for* Beginning Students

1

First Level

(Workbook)

Michael Angel C. G., Editor

Preface

Mathematical Olympiads for Beginning Students – The Zeroth Book for Elementary Schoolers is an initiative that arises at the suggestion of parents and teachers to want to count on a practical study material for the training of future Math Olympians who currently have nascent math skills.

This workbook covers a complete training program for beginners in five volumes, where each volume corresponds to a specific level of difficulty, and is especially aimed at elementary school children between 6 and 11 years old with little or no experience in Math Olympiads who seek to strengthen their math skills and become a Math Olympian. It may even be of great help for beginners in math, for whom it may be their first workbook on the subject. This volume includes the first level of the training program, which consists of a set of 10 exams, where each exam consists of 8 problems inspired by problems from Math Olympiads around the world. This workbook has been carefully designed so that the student can solve each of the problems in the same book without having to resort to additional sheets, thus having a complete and orderly record of all problems already solved. At the end of the book the student will find the answers to all the problems proposed in this volume.

It is worth mentioning that this series of workbooks is a sequential preparation material, that is, anyone who begins with this training program is recommended to start from the first volume without skipping any of them, in this way the students will experience the gradual improvement of their math skills, evidencing their progress continuously. Likewise, students are suggested to carry out the following training scheme: between 6 and 7 years old, up to volume 2; between 7 and 8 years, up to volume 3; between 8 and 9 years old, up to volume 4; between 9 and 11 years old, up to volume 5. It is important to clarify that what is suggested above is only referential as it is the minimum required for those ages; however, students are always encouraged to continually overcome themselves and face increasingly higher levels of difficulty.

For the success of this training program, the essential presence of a guide, tutor or parent is recommended during the learning process of the student, so that they can be guided in the face of doubts and encouraged in the face of obstacles that may arise. So students are encouraged to start their training as soon as possible and become a successful contestant in Math Olympiads, and parents are encouraged to ensure and closely monitor the proper preparation of their children.

Sincerely,

The editor

Contents

Problems

Exam 1

Problem 1. How many more cubes does Carl have than Bob?

Carl   Bob

Problem 2. In what figure did Jack draw Sonya correctly without losing any details?

(A) (B) (C) (D) (E)

Problem 3. Bob drew a castle of triangular and quadrangular shapes, shown in the figure. How many quadrangular shapes did he use?

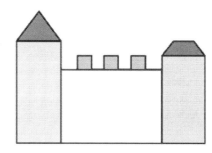

Problem 4. What figure will be obtained if black is replaced by white and white by black in the figure on the right?

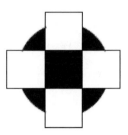

         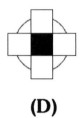   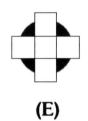

(A)   **(B)**   **(C)**   **(D)**   **(E)**

Problem 5. A grandmother divided a cherry pie into as many pieces as grandchildren. If there were 3 cherries on each piece of cake. How many grandchildren does the grandmother have?

Problem 6. In the figure, each number must be represented so that it is equal to the sum of the two numbers below it.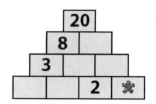

For example,
20	
8	12
. What number should be in the cell containing the symbol *?

Problem 7. In the figure, Luke colors all the squares, where the result is number 13. What pattern will he get?

8 + 4	19 – 6
20 – 5	6 + 7

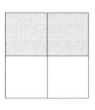   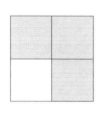   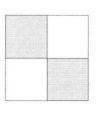   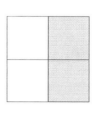   

(A)   (B)   (C)   (D)   (E)

Problem 8. Three foxes can carry the same load as a zebra. Two zebras can carry the same load as a giraffe. How many foxes does it take to carry the same load that a giraffe and a zebra can carry?

Exam 2

Problem 1. How many jumps does Jack need to do to reach his mother?

Problem 2. Which of the kangaroos is the tallest?

Joe   Nick   Ralp   Will   Jack

Problem 3. By correctly matching the puzzles shown in the figure, Johnny got three numbers that satisfy the addition. What number did he get as a result of this addition?

$$\boxed{} + \boxed{} = \boxed{}$$

2   1   7   5   5   4

Problem 4. The circle-shaped pizza was cut into 12 pieces. Chris ate two pieces, Joe ate one more piece than Chris, and Bob ate three more than Chris. How many pieces were left if no one else ate pizza?

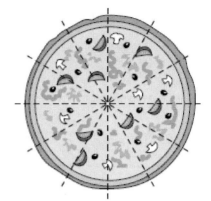

Problem 5. A dad has to change the wheels of five tricycles. Two wheels on the first and last tricycle, and three on all others. How many wheels does dad have to replace?

Problem 6. Lisa makes the numbers from the matches as shown in the figure below:

How many matches will she need to make the number 2018 according to this model?

Problem 7. Roxane baked some cakes. She ate a cake, then with Olenka they ate one each, then with Olenka and Vanessa they ate one each, and later with Olenka, Vanessa and her mother they also ate one each. If one cake left. How many cakes did Roxane bake?

Mathematical Olympiads for Beginning Students – *First Level*

Problem 8. There are several bees in a hive. 13 of them flew to collect pollen, but only 4 returned to the hive for lunch. There are now 21 bees in the hive. How many bees were in the hive at the beginning?

30

Exam 3

Problem 1. What frame shape is not used in the pictures?

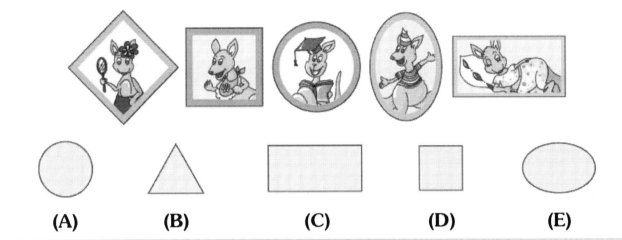

(A)   (B)   (C)   (D)   (E)

Problem 2. In a tree there were three crows, three cats, two titmice and Ivan. How many birds were on the tree?

Problem 3. Which of the options shows the correct weight of the basket of apples?

(A) 13 + 4   (B) 18 - 4   (C) 11 + 5

(D) 12 + 3   (E) 17 - 5

Problem 4. Jack and Robin represent different numbers in the equality shown. If Robin represents the greatest number. What number is Jack representing?

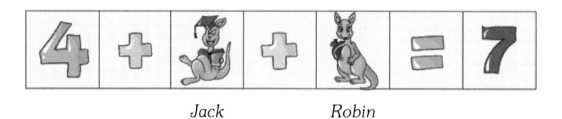

Jack   Robin

Problem 5. What is the shadow of the kangaroo shown in the figure opposite?

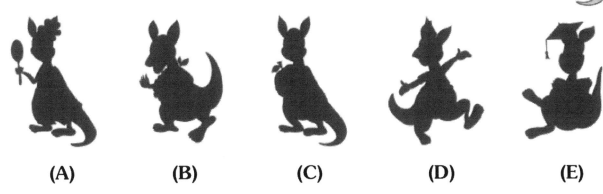

(A)   (B)   (C)   (D)   (E)

Problem 6. Fifteen children participate in a dance group, of which one is a boy and fourteen are girls. How many girls will participate in total after Sarah, Luke and Zoe have signed up for this group?

Problem 7. From which figure shown in the options can the house shown in the figure to the side be made?

(A)

(B)

(C)

(D)

(E)

Problem 8. There are two stories in a book. The first story occupies 11 pages of text and 3 pages of images. While in the second - 12 pages of text and 4 pages of images. How many total pages do the two stories take up?

Exam 4

Problem 1. Ralph has a flag with the number 10. Will wanted to compete with Ralph in math and wrote on some flags the sum of the numbers that he thought were equal to 10. Which flag is wrong?

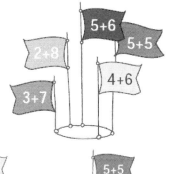

 3+7

 2+8

 5+6

 4+6

 5+5

(A)   (B)   (C)   (D)   (E)

Problem 2. The spider can move down, to the right, or to the left, but not up. If it must return at every intersection. What is the closest exit to the spider?

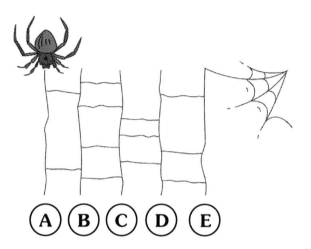

Problem 3. Help Kangaroo Jack to order the numbered cards from the largest to the smallest. Which is the right answer?

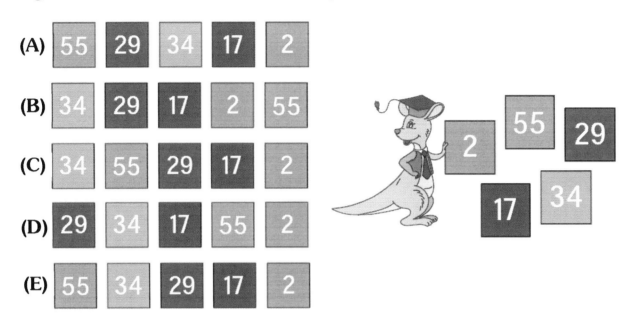

(A) 55 29 34 17 2

(B) 34 29 17 2 55

(C) 34 55 29 17 2

(D) 29 34 17 55 2

(E) 55 34 29 17 2

Problem 4. Two years ago, the sum of the ages of Bob and Rob was 15 years. Now Bob is 10 years old. In how many years will Rob be 11 years old?

Problem 5. What is the fewest matches that need to be added to the figure beside to get a square?

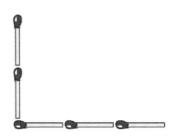

Problem 6. From the triangle shown in the figure to the side, cut 3 corners. How many angles does the new figure have?

Problem 7. Luke, Johnny and Bobby ate apples. Johnny ate 5 apples, Bobby - 3 apples. Luke and Johnny together ate three times the amount Bobby ate. How many apples did Luke eat?

Problem 8. Matthew and Joseph played foosball. The first match ended with a score of 2: 0 in favor of Matthew, the second match ended 2: 4 in favor of Joseph. Who won both games and with what score?

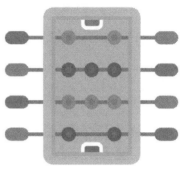

Exam 5

Problem 1. Each student told the teacher how many fairy tales they knew. The teacher wrote it on the board (see figure).

How many students knew less than 6 fairy tales?

Problem 2. From a 60 *cm* long loaf of bread, my mother cut a 15 *cm* long piece from each end for my twin brothers.

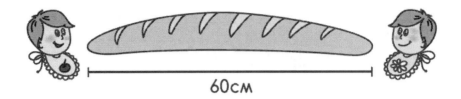

What is the length of the remaining part?

Problem 3. Tanya and Julia each received a bar of chocolate. Tanya has a whole bar and Julia ate part of it (see figure). How many squares of chocolate did Julia eat?

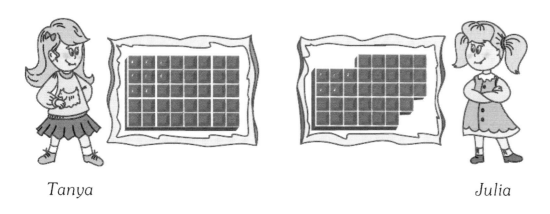

Tanya   Julia

Problem 4. There are 12 balls, 10 dice, and 7 small boxes in a large box. There is one candy in each small box. How many items (balls, cubes, boxes, and candies) are in two large boxes?

Problem 5. Nick wrote the numbers 1, 2, 3 and 4 in the cells of the 2 × 2 table (each number only once). He wrote the number 1 as shown in the figure. The sum of the numbers in the cells that have a common side with the cell in which the number 3 is written is equal to 5. What number did Nick write in the cell indicated by the question mark?

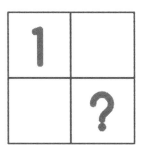

Problem 6. The letters K A N G A R O O are placed at some nodes on the grid shown in the figure. The side of the small square on this grid is 1 m. Go from letter P to letter Q along the grid lines and collect these letters in the specified order. What is the shortest length of that path?

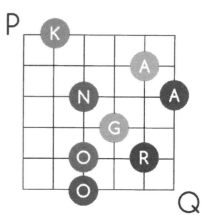

Problem 7. Jack places his photos in an album. If he places a photo on a page, he uses 4 stickers. If he places two photos on one page, he uses 6 stickers. How many stickers does Jack need to place on 3 pages with one photo per page and on 3 pages with two photos per page?

Problem 8. Albert has 5 dice. Two of the five are gray with white dots, the rest are white with black dots, and two of the five are small, the rest are large. Which figure shows Albert's cubes?

(A)  **(B)** **(C)**

(D) 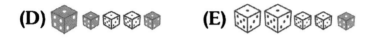 **(E)**

Exam 6

Problem 1. A mom made all-digit cookies for her children. Her youngest son Bob ate some cookies. The remaining cookies are shown in the figure opposite. Which cookies will Bob's siblings no longer be able to eat?

Problem 2. During the night, 3 kangaroos, 2 lion cubs, 1 beaver and 4 monkeys were born in the zoo. How many animals were born that night at the zoo?

Problem 3. Rick places two-colored balls on the steps of a staircase so that for each next step, he adds a ball, whose color differs from the color of the previous ball, that is, as shown in the figure. How will the balls be placed on the step marked with a question mark?

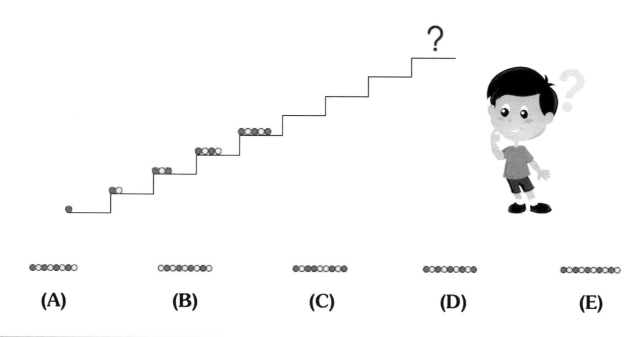

(A)   (B)   (C)   (D)   (E)

Problem 4. A group of 7 girls and 9 boys bought theater tickets in advance. Of these, 2 girls and 1 boy became ill and could not go to the theater. How many children in this group were able to attend the theater?

Problem 5. Only 5 digits were used to write the numbers 0, 1, 11, 12, 9, 8, and 19. What is the minimum number of numbers that need to be removed so that only three digits are used to write the remaining numbers?

Problem 6. My rabbit eats three carrots or a whole cabbage a day. It ate 12 carrots in a certain week. How many whole cabbages did my rabbit eat in that week?

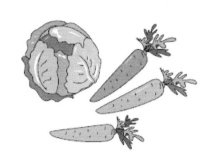

Problem 7. A bottle of milk costs 12 coins. The milk in this bottle costs 10 coins more than an empty bottle. How much does an empty bottle cost? (1 coin <> 100 cents)

Problem 8. One day three children met. Everyone greeted each other, and each greeting consisted of just two different words. How many words did these kangaroos say if it is known that no word was repeated twice?

Exam 7

Problem 1. A student solved five exercises, but made a mistake. Help the student discover in which exercise he was wrong.

(A) $20 + 1 - 7 = 14$ (B) $20 - 1 + 7 = 26$ (C) $2 + 0 + 17 = 19$

(D) $2 - 0 + 17 = 18$ (E) $20 - 1 - 7 = 12$

Problem 2. The figure shows 12 beads. Chris painted the first, third, and seventh bead in blue. The second and fifth in green. He did not paint others. How many beads were left unpainted?

Problem 3. Help Peter to determine how many balls are in the third box, if there are 2 more balls in the second box than in the first and 4 more than in the third.

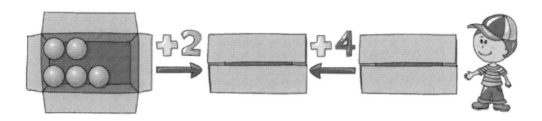

Problem 4. How many triangles and how many squares are shown in the figure?

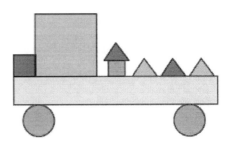

Problem 5. At the zoo, a rabbit eats 1 *kg* of cabbage, 2 *kg* of carrots and potatoes a day. If he eats potatoes and cabbage as much as carrots and cabbage. How many kilograms of potatoes does the rabbit eat per day?

Problem 6. Snail and Hedgehog's houses are located in different parts of the city from a fairy tale. The figure shows schematically the different paths between the houses as well as the distances between them in meters. What is the shortest distance between their houses?

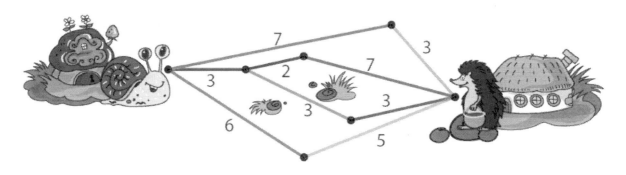

Problem 7. How many light gray squares need to be painted dark gray so that the number of light gray squares equals the number of dark gray squares?

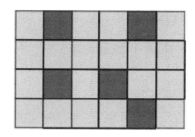

Problem 8. Which of the figures suggested in the options is NOT part of the landscape?

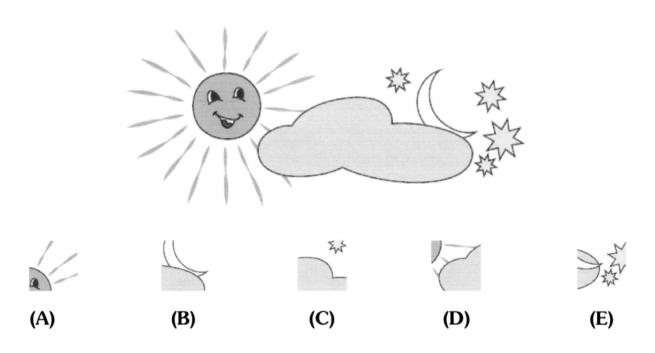

(A)   (B)   (C)   (D)   (E)

Exam 8

Problem 1. The figure shows circles and triangles. How many triangles are completely located (drawn) inside the circles?

Problem 2. How many times is the number 2018 written in the columns and rows of the table shown to the right?

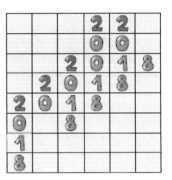

Problem 3. Mathew created a coding system to write the word MATEMATIKA using the numbers given in the table:

M	A	T	E	M	A	T	I	K	A
13	7	8	25	13	7	8	2	6	7

When he coded the word MIKE according to this system, he got the following set of numbers:

(A) 713713 **(B)** 132627 **(C)** 137287 **(D)** 825871 **(E)** 132625

Problem 4. In the land of adorable cats, a chocolate fish costs two cat coins and a candy mouse costs three cat coins.

Kitty Murzia bought three chocolate fish and a candy mouse. How many coins did she pay?

Problem 5. The figure shows seven paths in the park. Mary always starts her walk at point "O" and only goes through two paths. Which of the points in the park shown in the image could she never reach during the walk?

Problem 6. A cake was cut into 12 pieces. Bob and Matt ate 2 pieces each, Louis and John ate 3 pieces each. How many pieces are left?

Problem 7. Several shapes were removed from the pattern in Figure 1 and the pattern shown in Figure 2 was obtained. How many of those shapes were removed?

Figure 1

Figure 2

Problem 8. Kangaroo Jack wants to get the carrot by collecting as many coins as possible. He can jump to the next cell in each of the four possible directions, as shown in Figure 2. If the number of coins in each cell is shown in Figure 1. What is the largest number of coins Jack can collect in his journey to the carrot?

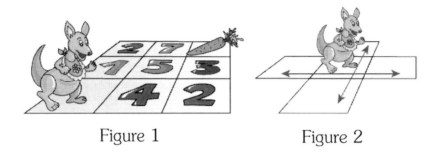

Figure 1   Figure 2

Exam 9

Problem 1. The snowman has five buttons, a pom pom hat, and a long nose. What image shows the snowman?

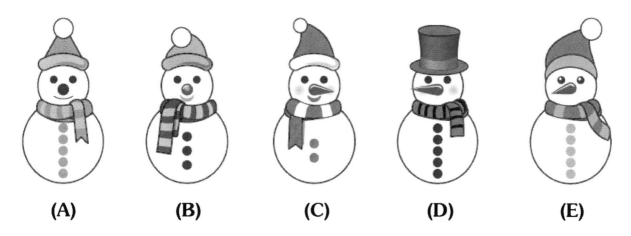

(A)   (B)   (C)   (D)   (E)

Problem 2. Anna classified the figures into three groups by color; see the figure below.

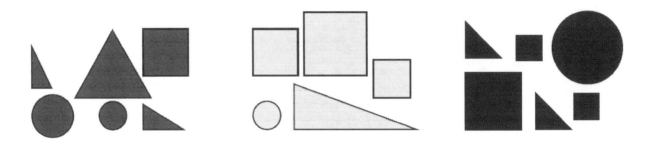

Peter re-classified the same shapes into three groups: triangles, squares, and circles. Which of his groups has the greatest number of shapes?

Problem 3. Albert plays a game, moving a chip as many positions as the die shows. He starts from cell 3 and rolls the die twice. If he first got 4, and then 5. To what cell number will Albert move his chip according to the rules of the game?

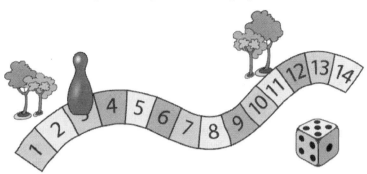

Problem 4. All the puppies in the figure weigh the same and the weights are in kilograms.

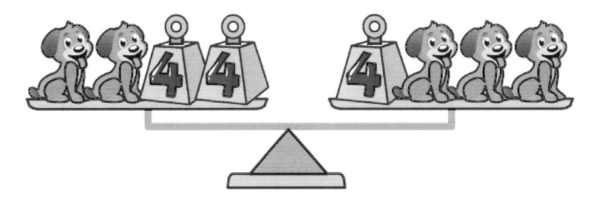

How much does a puppy weigh if the scale in the figure is balanced?

Problem 5. Sarah makes a necklace, stringing beads always in the same order (see figure). In what order should Sarah string four additional beads?

●■▲▭
1 2 3 4
(A)

▲■■▭
1 2 3 4
(B)

⬠▲▭■
1 2 3 4
(C)

●⬠▲■
1 2 3 4
(D)

⬠▲■■
1 2 3 4
(E)

Problem 6. Three daisies together have exactly 20 petals. Two of them are shown in the figure to the right. In which option can the image of the third daisy be seen?

(A)   **(B)**   **(C)**   **(D)**   **(E)**

Problem 7. Peter made a 9-cube arrangement on the floor. Then, he drew the top view of it. Which of the proposed figures is the arrangement made by Peter?

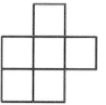

   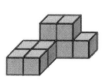

(A)   **(B)**   **(C)**   **(D)**   **(E)**

Problem 8. When you enter Grace's room through the door, you will see that there is a closet on your right and a window on your left. A chair to the right of the window and a shelf to the right of the closet. Which figure schematically shows Grace's room?

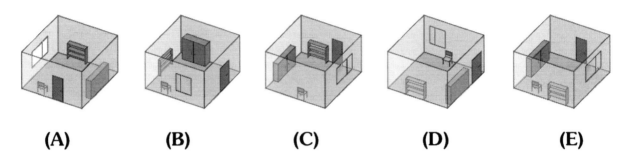

(A)   (B)   (C)   (D)   (E)

Exam 10

Problem 1. A box has balls of different colors: 2 red, 1 green, 3 blue, 2 yellow, and 1 purple. John drew 2 red, 1 green, 2 blue, 2 yellow and 1 purple from the box. What color is the ball that was left in the box?

Problem 2. A magician takes toys out of his hat in the same order:

What are the next two toys that the magician will take out from the hat?

(A)   **(B)**   **(C)**   **(D)**   **(E)**

Problem 3. The figure shows several pencils of the same size placed on the teacher's desk. How many pencils do not touch the desk?

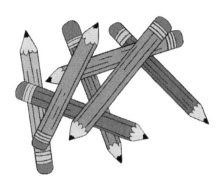

Problem 4. A beetle goes through a garland with flags. It started at the end marked with ∗ and is now between the two adjacent flags shown in the figure beside. How many flags will the beetle find on its way when it reaches the end of the garland without returning?

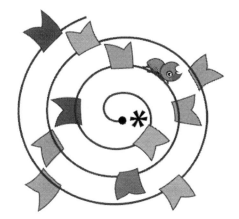

Problem 5. Which of the figures suggested in the options below is not a circle, is not a square and is not colored?

   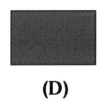 

(A)   **(B)**   **(C)**   **(D)**   **(E)**

Problem 6. How many sheep can be seen on the puzzle of four pieces when they are put together correctly?

 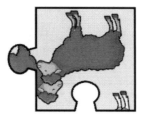  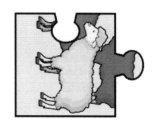

Problem 7. Identify a set that has only numbers greater than 4 and less than 10.

(A) 8,7,5,3   **(B)** 11,8,6,4   **(C)** 5,9,7,2   **(D)** 5,6,7,8   **(E)** 10,8,5,1

Problem 8. In the figure made of matches shown below, it can be seen 5 squares.

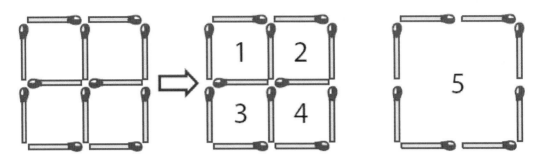

What is the least number of matches that have to be removed from this figure so that no squares can be seen in the resulting figure?

Answers

Exam 1

1. 2 cubes.

2. Option (C).

3. 7 quadrangular shapes.

4. Option (A).

5. 4 grandchildren.

6. 5.

7. Option (D).

8. 9 foxes.

Exam 2

1. 12 jumps.

2. Jack.

3. 75.

4. 2 pieces.

5. 13 wheels.

6. 20 matches.

7. 11 cakes.

8. 30 bees.

Exam 3	Exam 4
1. Option (B).	**1.** Option (C).
2. 5 birds.	**2.** Exit A.
3. Option (D).	**3.** Option (E).
4. 1.	**4.** In 2 years.
5. Option (C).	**5.** 7 matches.
6. 16 girls.	**6.** 6 angles.
7. Option (E).	**7.** 4 apples.
8. 30 pages.	**8.** Tie 4:4.

Exam 5	Exam 6
1. 6 students.	**1.** 6 and 9.
2. 30 *cm*.	**2.** 10 animals.
3. 6 squares of chocolate.	**3.** Option (D).
4. 74 items.	**4.** 13.
5. 4.	**5.** 2 numbers.
6. 24 m.	**6.** 3 cabbages.
7. 30 stickers.	**7.** 1 coin.
8. Option (C).	**8.** 12 words.

Exam 7	Exam 8
1. Option (D).	**1.** 2 triangles.
2. 7 beads.	**2.** 6 times.
3. 3 balls.	**3.** Option (E).
4. 4 triangles and 3 squares.	**4.** 9 coins.
5. 2 *kg*.	**5.** Point E.
6. 9 *m*.	**6.** 2 pieces.
7. 7 squares.	**7.** 11 shapes.
8. Option (E).	**8.** 24 coins.

Exam 9	**Exam 10**
1. Option (E).	**1.** Blue.
2. Squares.	**2.** Option (E).
3. 12.	**3.** 4 pencils.
4. 4 *kg*.	**4.** 8 flags.
5. Option (E).	**5.** Option (A).
6. Option (A).	**6.** 8 sheep.
7. Option (D).	**7.** Option (D).
8. Option (C).	**8.** 3 matches.

Answers

Made in the USA
Las Vegas, NV
03 February 2024

85247803R00068